# YOUR KNOWLEDGE HAS VALUE

- We will publish your bachelor's and
  master's thesis, essays and papers

- Your own eBook and book -
  sold worldwide in all relevant shops

- Earn money with each sale

Upload your text at www.GRIN.com
and publish for free

**Imprint:**

Copyright © 2016 GRIN Verlag, Open Publishing GmbH
Print and binding: Books on Demand GmbH, Norderstedt Germany
ISBN: 9783668354692

**This book at GRIN:**

http://www.grin.com/en/e-book/345362/global-health-and-climate-variation

**Francis Okpaleke**

# Global Health and Climate Variation

GRIN Publishing

Climate Change Evidence and Putative Risks for Global Health: A Desk Analysis

**Okpaleke Francis**

**Department of Political Science**

**Federal University Lafia**

**Abstract**

The climate is an integral component of planet earth supporting most life forms and processes (WHO, 2015). It is supports biotic and abiotic systems in the environment and provides cover to man from the ultraviolet emissions from the sun (McMichael, 2001). The world climate has however come under severe pressure over the last decades as a result of population explosion and increasing industrial and economic activities by man. As WHO (2005) puts it, 'climate change is man-made'. This man-made change to the earth's natural atmospheric balance has led climate scientists like (Pottier, 2014, Neer, 2004,) to predict that continuous emission of greenhouse gases would lead to a long term impact in the world's climate that would potentially affect human health. The aim of this paper is to outline the potential impact of climate change on global health. It provides evidence for the realization of these impacts while critically assessing what has been done and what can be done to mitigate further harm to human health.

Keywords: climate change, public health, evidence and risks.

# Content


Content ........................................................................................................................ 2

1. Potential Impact of Climate Change on Global Health ........................................... 3

2. Climate Change and Human Health ......................................................................... 6

3. Mitigating climate change risks? ............................................................................. 8

4. Possible Intervention strategies ............................................................................. 10

Conclusion ................................................................................................................. 11

References .................................................................................................................. 13

## 1. Potential Impact of Climate Change on Global Health

Anthropogenic activities of man have increased the presence of greenhouse gases (GHGs) in the earth's atmosphere over the last decades (Adeola, 1999). Emissions from Fossil fuel combustion, burning of forest, agricultural activities such as animal husbandry have increased the presence of carbon dioxide ($CO_2$) and methane in the atmosphere (Adeola, 1999). Synthetic halocarbons such as sulphate aerosol, chlorofluorocarbons (CFCs), and nitrous oxide have also been released into the atmosphere as a result of man-made activities (WHO, 2015)

As Vitousek (1999) observed, man-made activities is affecting the earth's biogeochemical systems. He notes;

> "...human alteration of the earth is substantial and growing since the industrial revolution. One third of the habitable land surface has been transformed by man, carbon dioxide concentration levels have increased by 30% and atmospheric nitrogen is fixed beyond natural thresholds. The earth's surface water has significantly depleted and one-quarter of the bird species are driven to extinction...by these standards, it is clear we live in a human dominated planet (Vitousek et al, 1999:19).

Climate change has been severally defined in literature. It refers to the alteration of the climate through man-made anthropogenic activities that alter the natural concentration of greenhouse gases in the earth's atmosphere (Briggs, 2002). The United Nations Framework Convention on Climate Change (UNFCCC) 2002 attribute climate change to:

"...directly or indirectly to human activities that alter the composition of global atmosphere and which in addition to natural climate variability observed over comparable time period" (International Strategy for Risk Reduction, 2008:2).

The risks of climate change to global health has been mostly neglected (Brown, 2003). Around the world, severe change in climatic conditions have led to the incidences of disease outbreak and increase human health risks (EPA, 2011). As Gareth notes, 'rising greenhouse gases deposition are engendering consequences that will affect human health in many ways' (Gareth, 2006). Episodes of high temperature, poor crop yield, changes in weather patterns, acidification, sea rise and desertification have been linked to mass migration, social and civil unrest which in turn leads to the spread of several infectious diseases (UNEP, 2003).

Potential impact of climate change on global health can have direct and indirect impacts. These effects ranges from environmental health issues, human psychological health, physical health and other infectious diseases (WHO, 2005). Direct impacts include rising temperatures which increases the incidence of extreme weather conditions like heat waves, heat stroke, skin cancer, asthmatic and heart conditions. Heat waves is a dangerous weather condition that have led to a number of deaths over the years. In Europe by 2003, heat wave was linked to the death of an estimated 22,000 to 35,000 persons (UNEP, 2003).

Direct impact of climate change has been predicted to cause extreme weather events like cyclones, storm-surges, droughts, and floods, whose aftermath increase production of certain air pollutants in the atmosphere and the transmission and spread of certain air-borne and water-borne infectious diseases like malaria and dengue fever especially in less developed climes (WHO, 2005). Irregular rain falls as a result of climate change resulting to drought would lead to the absence of clean water used for public sanitation, personnel hygiene and drinking, which can prompt an extensive variety of life-undermining disease such as diarrhea. Yearly, an estimated 760,000 children under the ages of five die from diarrhea (World Health Organization, 2015).

As Grant observes, rise in ambient temperature as a result of climate change has led to increases in the geographical distribution of particular vector organisms whose transmission and spread is enabled by certain biotic and abiotic factors including temperature, humidity, wind, vegetation and parasites (Grant, 2009). In warm climes like Africa, South America, climate change has been linked to the increase of mosquitoes which causes malaria and dengue fever, leishmaniasis (sand fly), schistosomiasis (water snail), yellow fever (mosquito), and other many vector borne diseases including encephalitis, and West Nile infection (Zhou, Minakawa, Githeko, & Yan, 2004). Malaria kills millions of people in Africa yearly more than HIV/AIDS (World Bank, 1992).

Studies also reveal that the first detectable change in human health as a result of climate change may be the alterations in geographical range and seasonal variations of vector-borne infectious diseases (Patz, 2002; Denis, 2004). Food borne infections like salmonella (salmonellosis) would increase during hotter weather conditions as a result of the production of noxious photochemical smog in urban areas.

Epidemiological studies by Folke et al (2004) have also established the link between increase in retention of heat energy in the lower atmosphere to incidence of skin cancer, eye lesions such as cataracts and immune suppression in fair skinned population especially in Europe, and North America. UNEP predicts that Europe and United States would experience an increase by 5-10% in the cases of skin cancer in the next five years and is expected to peak by 2020 as ambient ground level ultraviolet irradiation occasioned by the release of man-made ozone depleting gases (UNEP, 2011).

Indirect impact of climate change on human health would include increased demand for space, material and food which resultantly leads to decline in plants and animal population and worsen food insecurity crisis and malnutrition in certain regions especially in south Asia, Africa and central America where tens of millions are affected by increased desertification, seasonal disruption and crop failure (McMichael, 2002). Similarly public health systems would be affected as a result of the alteration in the natural food producing ecosystem, population displacement, civil strife and economic disruption engendered by climate change predicted (Watson, 2011).

## 2. Climate Change and Human Health

There is mounting evidence that the impact of climate change on human health would be realized. The UNFCCC 2012 report indicates that recent anthropogenic emissions are the highest in history as fossil fuels and industrial production have accounted for 78% of all emissions since 1970 (UNFCCC, 2012). Emerging evidences already show melting ice caps, ocean acidification, global temperature rise, sea level rise and increase in diseases due to climate change (WHO, 2015). This is further supported by several findings made by climate scientists and meteorologists through model simulations and climatic observations which envisages worsening human health throughout the 21$^{st}$ century especially in developing countries with low incomes and limited capacity to respond to climate change baselines (Cubasch, 2001, Martens, 2006).

Already evident impact of climate change to human health in regions is documented in literature. In Europe, climate change has been linked to the increasing levels of thermal stress and air pollution as well as vector and water borne diseases (Shindell, 2001). The resurgence of malaria in south Europe, especially in Italy and Albania have been linked to rise in temperature (WHO, 2008). In Australia and New Zealand, climate change has been linked to increase in cases of mortality as a result of heat waves, heat strokes and severity of natural disasters like floods and storms (Epstein, 2011). In Australia, arbovirus infections and the Aedes mosquito have increased over the years with reported cases of encephalitis and Ross River virus (Langford, 1995).

In many parts of Africa, climate change has been linked to increase in endemic diseases including malaria, rift valley fever, yellow fever, meningitis, cholera and dengue fever. Increased desertification has increased the food insecurity in the region with a millions in Africa suffering from malnutrition and food borne diseases (WHO, 2015). In Asia and Latin America, just like Africa, tropical infections like malaria, dengue fever, plague, leishmaniasis, cholera and Chagas diseases are reported (WHO, 2015)

### 3. Mitigating climate change risks?

Several actions have been adopted to mitigate the impact of climate change over the years with many countries establishing clear regulations to reduce emissions.

The UNFCCC (United Nations Framework Convention on Climate Change) is an international forum created in 1992 to tackle the threat of climate change with over 195 countries as signatories to the UNFCCC agreement as at 2010 (UNFCCC, 2010). The framework focuses on four key elements of climate change control: adaptation, financing, mitigation and technological development.

Adaptation aims at reducing substantial GHGs emissions over the next few years and adapting to the threat of climate change by reducing vulnerability and increasing resilience of developing nations (Cancun Adaptation Framework, 2010). Mitigation like adaptation focuses on reducing emissions and achieving a near zero carbon dioxide emission and other GHGs by the end of the century. This it expects to be achieved by proper financing of these pathways and encouraging states to adopt more eco-friendly and greener technologies (UNFCCC, 2012).

So far, 37 industrialized countries including the United Kingdom has set commitment to reduce emissions of GHGs under the Kyoto protocol in 1997 and have been encouraged to further reduce emissions under the UNFCCC framework. The United Kingdom for example, initially started with a 12.5% reduction commitment of GHGs and further increased its GHG reduction commitment to 27% in 2011 (World Climate, 2012).

The Kyoto protocol treaty extends from the UNFCCC, it commits member states to reduce GHGs based on the effect of global warming and anthropogenic activities of man (Kyoto protocol, 1997). It sets benchmarks for countries to reduce emissions of GHGs. The first Kyoto protocol set a 5% reduction for the first commitment period (2008-2010), while these have been remarkably achieved, it has however not been sufficient to reduce the increasing emissions from other countries not party to the protocol.

The second Kyoto protocol agreed for the period 2013-2020 saw fewer countries participating in it with many countries making voluntary pledges to cut emissions. In 2015, the UNFCCC agreed to develop a new protocol legal instrument with binding legal force applicable to all parties. Countries have been made to recognize the need for urgent GHG reductions and the need to keep global temperatures within acceptable pre-industrial levels (World Climate, 2012).

The Synthesis Report (SYR) which details efforts of working groups on climate change is also part of the efforts by IPCC (Intergovernmental Panel on Climate Change) to provide a comprehensive report on climate change using current scientific and technical assessments to document climate change. The release of the SYR in 2012 in Copenhagen was instrumental to the 20th Conference of Parties in Lima and the 21st session in 2015 in Paris under the UNFCCC to develop unilateral action for the reduction of GHGs (IPCC, 2012).

## 4. Possible Intervention strategies

The following are suggested mitigation strategies to reduce the possible risks of climate change on human health:

Firstly as matter of policy action and preserving the world climate, countries should be legally bound to commitments aimed at reducing the emission of GHGs if the impact on human health is to be mitigated. This would require more holistic adaptation planning and implementation to be embraced by national governments as well providing adequate finance especially for developing countries to boost capacity for adaptation including developing greener technologies for industrial processes.

Furthermore, environmental sustainability planning and management practices would also help to reduce global warming. Silviculture, forest management, afforestation, reclamation of desertified areas, sustainable lumbering and responsible sourcing of ecosystem goods and services would de-carbonize the atmosphere while providing viability of microbiota and habitats for wildlife. This would also help curb the rising incidence of heat stroke, heat stress, skin cancer and lesions as a result of rising temperatures. This will also possibly help reduce the global disease burden.

As part of measures to ensure adaptation and mitigation to climate health risks, government at all levels should increase surveillance and response to infectious diseases. As climate change has been established to affect water quality and increase the likelihood of contamination and water-borne diseases, pro-active measures would entail providing cleaner water sources and expedite actions to control the spread of infectious diseases. Proper vaccination would also help to ameliorate the long term effects of these diseases.

Disaster management is a pro-active intervention strategy that would help improve public health and environmental health planning in the case of natural disasters such as cyclones, storm surges and floods which increases the risks of infectious diseases after such events. Key to minimizing health risks would entail building capacity for health services to adequately respond to emergencies and establishing a national disaster management agency where lacking to support national health actors.

Climate change is a global issue so are the health risks. Effective mitigation strategies would require stronger enforcement of international regulations limiting global emissions of GHGs. The UNFCCC and the Kyoto protocol should be binding and countries that produce more GHGs should made to conform to global quota for emissions. To this end, international organizations like the United Nations Environmental Programme (UNEP), World Meteorological Organizations should work with national governments to develop better and appropriate safeguard mechanisms for adapting and mitigating the impacts of climate change to health.

**Conclusion**

From the foregoing it has been established that climate change engenders conditions that have severe health risks to humans. Extreme weather conditions and disruption in weather patterns have increased the incidence of skin cancer and heat related diseases. Similarly, several water-borne, air-borne and vector-borne diseases such as malaria, dengue fever have been linked to the effects of climate change.

Rapid changes in demography, population, and economy occasioned by the industrial revolution over the decades and by man's anthropogenic activities and quest to conquer nature have increased the risk to its health and the environment as more and more GHGs are deposited into the atmosphere. The task lies in developing better strategies to adapt and mitigate the impact of these changes while taking proactive measures to address deposition of these gases in the atmosphere if mankind is to survive or be annihilated over the next decades.

**References**

Adeola, F.O. (1999). Endangered community, enduring people: Toxic contamination, health, and adaptive responses in a local context. *Environment and Behavior* 32, 2, 207-247.

Epstein , P.R (2011). Climate and health. *Science* 8(1)

Folke, C et al (2004). Renewable resource appropriation In: Getting Down to earth. Washington DC, USA, Island Press. Pp 11

Gareth U., (2006) Health effect of climate change in England [online] Available athttp://www.gov.uk/government/uploads/system/uploads/attachment_da ta/file/371103/Health_Effects_of_Climate_Change_in_the_UK_2012_V1 3_with_cover_accessible.pdf [Accessed 10 Apr. 2016].

Intergovernmental Panel on Climate Change (IPCC). *Climate change 2001: Third Assessment Report (volume II), Impacts, Vulnerability and Adaptation.* Cambridge: Cambridge University Press, 2001

International strategy for risk reduction. (2008). *climate change and disaster risk reduction.* [online] Available at: https://www.wmo.int/pages/prog/dra/vcp/documents/7607_Climate-Change-DRR.pdf [Accessed 4 May. 2016].

Langford, I. H (2001). The potential impact of climate change on winter mortality in Europe. *International Journal of Biometeorology* 38:141

Jeremy M., (2003). *Human health effects of air pollution.* [online] Sciencedirect.com. Available at: http://www.sciencedirect.com/science/article/pii.html [Accessed 10 May 2016].

Martens, W.J.M., (2006) *Climate change and future population at risk of malaria. Global Environmental Change* 9 (1) pp 102.

McMicheals, A. J (2002). Human frontiers, environment and disease. Cambridge, UK, Cambridge University Press.

Neer B., (2014). *Essentials of environmental public health.* London: University Press

Patz, J.A (2002) Global climate change and emerging infectious diseases. *Journal of the American Medical Association* 275: pp 217

Portier, C.J., (2013). A human health perspective on climate change: a report outlining the research needs on the human health effects of climate change. *Journal of Current Issues in Globalization, 6*(4), p.621.

Shindell, D.T (2001). Increased polar stratospheric ozone losses and delayed eventual recovery owing to increasing greenhouse gas concentrations. *Nature* 392: 589.

UNEP, United Nations Environment Programme (2011) The Stockholm Convention on Persistent Organic Pollutants (POPs). http://www.pops.int (accessed 2nd February 2016).

United Nations Framework for Climate Change (2002) available online http://www.unfcc.org (accessed 5th May 2016)

United States Environmental Protection Agency (EPA) (2011). Greenhouse effect schematic. http://www.epa.gov/air/aqtrnd95/globalwarming.html (Accessed 9th May, 2016).

Vitosuek, P. M et al (1997). Human domination of Earth's ecosystems. Science 277: 494-499

Watson , P.R., (2001). Global climate change-the latest assessment: Does global warming warrant a health warning? *Global Change and Human Health* (2)1

WHO, World Health Organization (2005) Polynuclear aromatic hydrocarbons in drinking water. Background document for development of WHO guidelines for drinking-water quality. http://www.who.int/water_sanitation_health/dwq/chemicals/polyaromahyd rocarbons.pdf (accessed 2nd May 2016).

World Bank (1992). World development report. Development and the environment. Oxford, UK, Oxford University Press.

World Climate (2014). A balanced response to the risks of dangerous climate change (online) http://www.theccc.org.uk/tackling-climate-change (accessed 10th May, 2016).

World Health Organization (2009). Impact of climate change on communicable diseases. Geneva [online] Available at: http://www.searo.who.int/EN/Section10/Section2537_14458.html [Accessed 7 May 2016].

World Health Organization. (2015). *Climate change and health.* [online] Available at: http://www.who.int/mediacentre/factsheets/fs266/en/ [Accessed 3 May. 2016].

Zhou, S. Minakawa, J., Githoeko, W, Yan J., (2004). Climate change effects on human health: projections of temperature-related mortality for the UK during the 2020s, 2050s and 2080s. *Journal of Epidemiology & Community Health*, 68(7), pp.641-648.